Amanda Costa
Leocádia Beltrame

Feasibility of Composting in Urban Solid Waste Management

Amanda Costa
Leocádia Beltrame

Feasibility of Composting in Urban Solid Waste Management

Case study in Paulista-PE

ScienciaScripts

Cover image: www.ingimage.com

This book is a translation from the original published under ISBN 978-3-330-77206-9.

Publisher:
Sciencia Scripts
is a trademark of
Dodo Books Indian Ocean Ltd. and OmniScriptum S.R.L publishing group

120 High Road, East Finchley, London, N2 9ED, United Kingdom
Str. Armeneasca 28/1, office 1, Chisinau MD-2012, Republic of Moldova, Europe
Managing Directors: Ieva Konstantinova, Victoria Ursu
info@omniscriptum.com

Printed at: see last page
ISBN: 978-620-8-64086-6

SUMMARY

SUMMARY

When improperly disposed of, solid waste is an environmental problem, as it is harmful to ecosystems and human health. One of the techniques for treating solid waste is composting, which has the advantage of producing fertilizing compost at the end. In addition, it promotes the treatment of the organic fraction, which corresponds to the majority of urban waste, reducing the amount sent to landfills and increasing their useful life. In view of the above, the aim of this study is to analyze the potential of this form of solid waste treatment in the municipality of Paulista, located in the Recife Metropolitan Region, which produces 11,800 tons of MSW per month, 62.7% of which is organic waste. Due to the fact that the municipality does not have selective collection, the feasibility of implementing composting for pruning waste, which is collected separately and produces 200 tons per month, will be studied. The proposed method is natural, as 7 tons per day will be treated. The municipality has great potential for including this treatment in the management of its waste, as it has the area available for operation and the technique is already provided for in the Waste Management Plan, and would represent a reduction in costs for municipal management.

Key words: Waste management, Organic waste, Pruning waste.

1 INTRODUCTION

The environment is one of the main issues under discussion worldwide due to the effects of human action on nature. Included in this issue is the proper treatment and final disposal of the waste produced by human activities, since when these are not disposed of properly, they affect both the environment and human health. Population growth, accelerated urbanization and industrial development have increased the pressure on natural resources and, consequently, the production of solid waste.

Their inadequate management can lead to pollution of water sources, air, soil degradation, an increase in the incidence of diseases caused by vectors that proliferate with the accumulation of waste and even damage to tourism (COUTINHO et al., 2011).

In order to promote the principles and guidelines relating to the integrated management of solid waste, the National Solid Waste Policy (PNRS) was instituted in Brazil on August 2, 2010, through Law 12.305 (BRASIL, 2010). According to the Brazilian Standard (NBR 10.004:2004), solid waste is

> ...solid and semi-solid waste resulting from industrial, domestic, hospital, commercial, agricultural, service and sweeping activities. Included in this definition are sludges from water treatment systems, those generated in pollution control equipment and installations, as well as certain liquids whose particularities make it unfeasible to discharge them into the public sewage system or bodies of water, or require solutions that are technically and economically unfeasible in view of the best available technology (ABNT, 2004, p. 1).

There are six main ways of disposing of solid waste: the "dump", controlled landfill, sanitary landfill, incineration, pyrolysis and composting. According to Philippi Jr (2005) apud Coutinho et al. (2011), in order to achieve a solution to the problem of waste in a way that is economically, environmentally and socially sustainable, various alternatives must be used in the management of solid waste. Considering, therefore, the link between measures to reduce generation at source and various methods of treatment and disposal. Composting is considered an environmentally appropriate final destination for waste, according to the definitions of the PNRS, which also includes reuse, recycling, recovery and energy use (BRASIL, 2010).

The municipal administration is responsible for the management of solid urban waste (MSW) (POLAZ; TEIXEIRA, 2009), excluding the responsibility of its generators, and has a legal obligation, established by the National Solid Waste Policy, to extinguish open dumps (BRASIL, 2010). MSW is generated in the urban environment, consisting of materials from households, commercial establishments and sweeping and cleaning services in public places, which are the exclusive responsibility of the municipality, from collection to final disposal.

Setting up landfills is still a problem due to the high costs and scarcity of available and suitable areas, as well as the environmental degradation. It is therefore urgent and necessary to expand the use of other technologies for treating solid waste (COUTINHO et al., 2011).

2. THEORETICAL FRAMEWORK

In order to articulate scientific knowledge with the study developed, the theoretical framework used will deal with the composting process and the management of solid urban waste in Brazil.

2.1 Ways of treating solid waste urban

The so-called "dump" is an inadequate form of disposal, characterized by dumping on the ground without criteria and environmental protection measures, causing public health problems such as the proliferation of disease vectors and pollution of the soil and surface and groundwater by leachate, a liquid produced by the decomposition of the organic matter contained in the waste (FEAM, 2006).

The controlled landfill, according to the Brazilian standard (NBR 8849:1985) (ABNT, 1985) is a technique that does not cause damage or risks to public health and safety, minimizing environmental impacts, using engineering principles to confine solid waste, covering it with inert material. However, this method still causes localized pollution, as there is no treatment of the leachate, no extraction or controlled burning of the gases produced, and no sealing of the base, which causes contamination of the soil and groundwater (FEAM, 2006).

The sanitary landfill differs from the controlled landfill in that it confines the solid waste to the smallest possible area and the smallest permissible volume, covering it with layers of earth at the end of the working day, or if necessary at shorter intervals (ABNT, 1992). In order for this technique to be environmentally suitable, there are technical structural requirements to follow, as well as a prior study of the potential impact on the area of influence of the project. The environmental protection elements relating to the structure of the landfill include a waterproofing system for the base and sides; daily covering and final covering; collection and drainage of percolated liquids; collection and treatment of gases; surface drainage system; treatment of percolated liquids and monitoring (FEAM, 2006).

Incineration consists of burning materials for a predetermined period of time at a high temperature, usually above 900°C, with the appropriate mixture of air (CAIXETA, 2005). It is the form of treatment commonly used for contaminating or hazardous waste, such as hospital waste, because the process of thermal destruction through controlled combustion reduces the weight and volume of the waste and its hazardous and pathogenic characteristics. In addition, the heat can be used to generate energy (COUTINHO et al., 2011).

Pyrolysis consists of the thermal degradation of hydrocarbons in the absence of oxygen. It requires an external source of heat to be supplied to the material, so the temperature can vary between 300°C and 1000°C (RODRIGUES et al., 2014). Through pyrolysis, organic matter can be converted into various by-products, including gases, composed of hydrogen, methane and carbon monoxide; liquid fuel and a solid residue, basically made up of almost pure carbon and other inert materials. It allows energy to be used through the thermal decomposition of waste in a controlled atmosphere. However, it is a process that is still under development as far as the treatment of solid urban waste is concerned, and therefore requires more rigorous studies (AIRES et al., 2003).

Composting is the process of biological decomposition and stabilization of organic substrates through the action of different microorganisms. It is related to the management of organic material by man, who has observed natural processes and developed techniques to speed up decomposition and produce the necessary organic compounds. The waste used can be of urban, industrial, agricultural or forestry origin (CERRI, 2008). In general terms, the composting methodology consists of choosing raw materials that offer a carbon/nitrogen balance favorable to the metabolism of the organisms that will carry out the degradation; facilitating the degradation by choosing the right location, according to the type of fermentation; controlling humidity, aeration, temperature and other factors (KIEHL, 1985).

2.2 The composting process

Composting is a controlled aerobic process in which organic matter is transformed through physical, chemical and biological processes, carried out in two distinct phases. The first, biostabilization or semi-maturation, when the most intense biochemical reactions occur, predominantly thermophilic, and pathogenic bacteria are eliminated; the second phase, maturation, is when humification occurs (GOUVEIA, 2012). Composting time will depend on the technology used and the type of waste to be composted, but generally ranges from 25 to 35 days for the first phase and 30 to 60 days for the second phase (CAMPOS; BLUNDI, 1998).

The use of organic, vegetable and animal waste has been adopted by farmers since ancient times to encourage plant growth and improve agricultural production. The oldest records of the use of compost in agriculture appear on clay tablets in the Mesopotamian Valley, 1000 years before Moses. The Romans and Greeks were already familiar with composting and comments about compost can be found in medieval religious texts and Renaissance literature. The Chinese also applied the principles of composting at this time (PIRES, 2011).

Compost was initially made without specialized techniques. In 1843, George Bommer filed a patent in America for a process that placed agricultural waste on a grid for decomposition, placing the run-off on top of the pile to speed up the process. After 15 days, the product already had the characteristics to be incorporated into the soil as an organic fertilizer. Although simple, the method is considered a pioneer in scientific composting (KIEHL, 1985; CORDEIRO, 2010).

At the beginning of the 20th century, the English phytopathologist Sir Albert Howard developed a technique for producing fertilizers by observing the empirical methods used by the natives of Indore, India. His technique became known worldwide as the Indore method or Howard method, used especially for agricultural waste (KIEHL, 1985). Between 1926 and 1940, Waksmam and his associates complemented their large-scale experiments with laboratory-scale studies using animal manure mixed with plant remains. These studies, as well as those of other researchers at the time, made it possible to establish the influence of certain factors on the composting process, such as humidity, aeration and temperature, as well as the carbon/nitrogen ratio and pH (KIEHL, 1985; CORDEIRO, 2010).

In Brazil, Dafert, the first director of the Agronomic Institute of Campinas, was the first to encourage farmers, in activity reports between 1888 and 1893, to produce fertilizers on their properties that were classified as "national manures" because the mineral fertilizers of the time were imported. In 1945, Aloisi Sobrinho, from the same institution, published a paper on a technique which consisted of preparing a broth with water and various animal manures to inoculate the plant remains to be composted. In 1950, the Luiz Queiroz College of Agriculture also began to encourage the preparation of compost (KIEHL, 1985).

According to Cunha Queda (1999) apud Cordeiro (2010), three fundamental stages can be identified in the composting process, which are the conditioning of the materials, the composting process itself, and the thinning of the compost.

and refining the compost. Conditioning the materials refers to the pre-processing stage in which the organic fraction is separated from the other materials and the particle size is reduced to facilitate the action of the microorganisms and maintain ideal conditions for aeration. At this stage, agents can also be added to adjust the mixture. The adjustment can be of a structural nature, in which support agents are mixed to ensure a good structure of the mixture, or conditioning, which has the function of correcting the carbon/nitrogen ratio or the moisture content (CORDEIRO, 2010).

Composting itself is the oxidation of the selected organic fraction due to the attack of various microorganisms under aerobic conditions. Therefore, the material must be turned periodically or receive air through aerators. Two main phases are considered: the active phase, characterized by high temperatures, intense decomposition reactions, the release of heat, CO_2 and water vapour; and the finishing phase, in which the temperature returns to equilibrium with the ambient temperature, the organisms reach a dynamic equilibrium and there is synthesis of human substances. Refining the compost corresponds to the mechanical post-processing phase with the aim of improving the granulometric characteristics of the compost and removing inert contaminants that were not removed in the first stage (CERRI, 2008; CORDEIRO, 2010).

As the composting process involves the action of microorganisms on organic matter, the factors that affect this process are those that directly or indirectly influence the varied population of organisms needed for decomposition. Thus, oxygen will supply the biological demand; temperature will affect the speed of biochemical reactions; and humidity, without which metabolic activities will not take place (KIEHL, 1985; RUSSO, 2003). The composting process is relatively insensitive to pH values, as organic matter with a pH between 3 and 11 can be composted, with values close to neutrality (5.5 to 8) being considered ideal (CERRI, 2008). In addition to these factors, an adequate C/N ratio must be considered so that the microorganisms can act. Another factor that interferes with the process is particle size, i.e. granulometry, which influences aeration and the geometric stability of the piles or beds (CORDEIRO, 2010).

2.3 Factors affecting the composting process

2.3.1 *Aeration*

Aeration of the compost is essential for the process to be efficient, as aerobic microorganisms need oxygen to carry out their metabolism. With plenty of air, decomposition is faster and more controlled, avoiding excess temperature and humidity, as well as bad smells. Compost heaps can be aerated by manual or mechanical turning, causing the outer layers to occupy the inner part. The supply of oxygen can also be done by air insufflation (KIEHL, 1985).

2.3.2 *Temperature*

High temperatures are necessary for composting. However, an optimum development temperature must be maintained for the microorganisms, as a variation upwards or downwards causes a reduction in population and metabolic activity (Figure 1) (RUSSO, 2003). Kiehl (1985) considers the optimum range for composting to be 50 to 70°C, with 60°C being the most suitable. Temperatures above 70 degrees are considered unnecessary and even inadvisable for long periods, as they limit the number of microorganisms that can live there. Up to a certain value, the temperature influences the speed of degradation of organic matter, after which it is harmful because it eliminates them. Excess temperature can also prevent the action of enzymes, slowing down the activity of microorganisms and, consequently, the composting process (RUSSO, 2003).

Figure 1. Temperature variation throughout the composting process

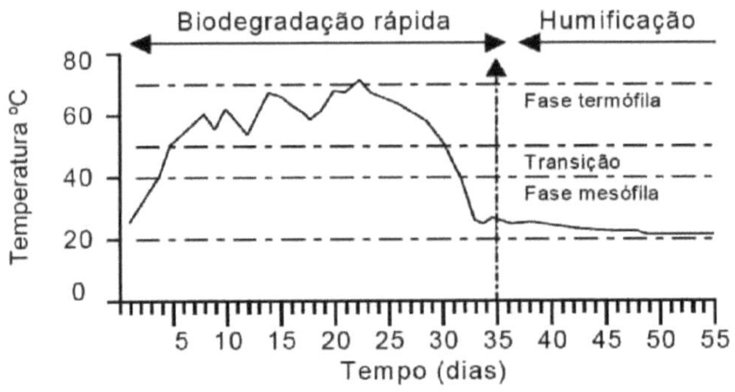

Source: Pires (2011)

2.3.3 *Humidity*

As composting is a biological process for the degradation of organic matter, the presence of water is essential for the physiological needs of the microorganisms. Theoretically, the moisture content should be 100%, but it varies throughout the process due to various parameters, such as: type of organic matter, particle size, geometric configuration of the composting bed, specific weight of the composting mass, aeration system and form, among others (PEREIRA NETO; LELIS, 1999; RUSSO, 2003). According to a study carried out by Pereira Neto and Lelis (1999), at the start of the process the ideal level is 60%, if it is below 30% it is harmful as it inhibits microbiological activity, and during the process the humidity cannot fall below 40%. High humidity levels promote the occupation of empty space with water, restricting the occupation of air and the diffusion of oxygen.

2.3.4 Carbon/Nitrogen Ratio

The appropriate ratio of carbon and nitrogen contributes to the growth and activity of the colonies of microorganisms involved in the process of degrading organic matter, making it possible to produce compost in less time (CERRI, 2008). According to Russo (2003, p. 57):

> Carbon has three main physiological functions: a) it is a constituent of cellular material; b) it functions as a donor electron in energy metabolisms (respiration of organic substrates and fermentation); c) it functions as a receptor electron in energy metabolic reactions (fermentation, reduction of CO_2 into CH_4). For its part, nitrogen: a) is a constituent of proteins, nucleic acid, coenzymes and amino acids; b) acts as a donor electron in the energy metabolic reactions of certain bacteria; c) in the form of nitrite and nitrate, it acts as a receptor electron in the energy metabolic reactions of denitrification bacteria under anaerobic conditions.

According to Kiehl (1985), these microorganisms absorb carbon and nitrogen in a ratio of 30 parts of the former to one part of the latter, which is therefore the ideal ratio in waste. However, the limits of 26/1 to 35/1 are considered to be the most recommended C/N ratios for fast and efficient composting. If the waste used for composting has a low C/N ratio, it loses nitrogen in ammoniacal form, damaging the quality of the compost. In this case, it is advisable to add cellulosic vegetable waste, such as pruning and gardening waste, to bring the C/N ratio close to ideal. If the raw material has a high C/N ratio, the process becomes slow and the final product will have low levels of organic matter. To correct this, simply add nitrogen-rich materials such as manure, animal bedding, vegetable pies, etc. (RUSSO, 2003; CORDEIRO, 2010).

2.3.5 pH

The composting process is relatively insensitive to pH values, as organic matter with a pH between 3 and 11 can be composted. However, values close to neutrality are considered ideal (5.5 to 8) because this is where microorganisms adapt best. At the start of the process, the pH reaches low values, close to 5, due to the action of the bacteria, and over the course of the process, as the compost stabilizes, it reaches values between 7 and 8 (CERRI, 2008).

2.3.6 Material granulometry

Kiehl (1985) considers the size of the particles to be a fundamental characteristic for the smooth running of composting, as their size will define the surface area exposed to attack by microorganisms. The larger the exposure surface, the smaller the particle size of the organic matter and the faster it will decompose. However, granulometries that are too fine, less than 2 mm, hinder aeration by facilitating compaction, while those above 16 mm facilitate natural aeration (RUSSO, 2003). Therefore, a value must be found that takes into account aeration and the exposure surface. According to Cordeiro (2010), the size of the particles used in composting cannot exceed 3 cm in diameter.

2.4 Composting methods

Composting can be done in an open or closed environment. Open environments are those in which the mass to be composted is placed in heaps, or beds, in composting yards. Closed environment processes are those that take place in digesters in the form of rotating drums, tanks, silos or cells, all of which have mechanical revolvers to move the organic mass. Composting in an open environment requires a larger area than in a closed environment, and the curing time is also longer. The two types of process can be combined, as shown in Table 1 (KIEHL, 1985).

Table 1. Classification of composting methods

Open Systems	**Closed systems**
Revolving Battery	Vertical reactors: - Keep going; - Discontinuous.
Static Battery: - Air intake; - Air inflation; - Alternating ventilation (suction and insufflation) - Air insufflation with temperature control.	Horizontal reactors: - Aesthetics - With material handling.

Source: Adapted from Cordeiro (2010)

The composting process can be static or slow, which is natural composting, which takes place passively and the period for obtaining compost exceeds 100 days; or fast or dynamic, which consists of an accelerated composting process in which special conditions are offered, such as the addition of enzymes and forced aeration (PIRES, 2011).

Natural composting consists of forming piles (Figure 2) with the organic matter to be decomposed, approximately 1 m high and 2 m wide at the base, for manual turning, and 2 m high and 4 m wide at the base for mechanical turning. The mixture should have a Carbon/Nitrogen ratio of between 25:1 and 35:1. Due to the reactions resulting from the metabolism of the microorganisms, the temperature rises to around 70°C in a few days. For an adequate supply of oxygen, the compost should be turned every three days, after which the temperature will reach 70°C again, and so on until the 70th day, when the material will be semi-cured and the temperature will have stabilized close to room temperature. In this way, the compost will be ready for use between 90 and 120 days. This method presents some difficulties, such as the need for large areas for the allocation of the beds and turning, as well as the environmental impacts resulting from the formation of percolated liquids, leachate, the production of gases, the generation of odors and the proliferation of vectors, which are complicated to control (PIRES, 2011).

Figure 2: Piles in a compost yard

Source: MMA (2010)

Accelerated composting can be carried out using static beds with forced aeration, in which various types of organic waste can be used, such as sewage sludge, household waste and vegetable pruning waste. The aeration of static beds is carried out by a system of perforated tubes for aeration or exhaustion on which the organic fraction to be decomposed is deposited. A pile can be 2 to 2.5 m high and is usually covered with a layer of cured and sieved compost to reduce characteristic odors. Each pile has an individual blower or exhaust fan to better control aeration. The composting time is three to four weeks, and then another four to five weeks for the material to cure (PIRES, 2011).

In addition to this technique, there is accelerated composting using reactors, in which, in addition to forced aeration, there is a closed shelter containing the organic material to be composted. A variety of shapes can be used as reactors in this system: vertical towers, horizontal towers (rectangular or circular), and circular rotating tanks (Figure 3). In this process, the generation of percolated liquids and the exhalation of gases can be better controlled, as can the control of humidity and temperature, which is done through precise instruments and forced aeration, reducing the area used by at least 50%. The fundamental characteristic of this type of system is the extreme reduction in composting time, from 120 days to 30 days, while the others are subject to climatic changes (op. cit.).

Figure 3: The most common reactors in composting

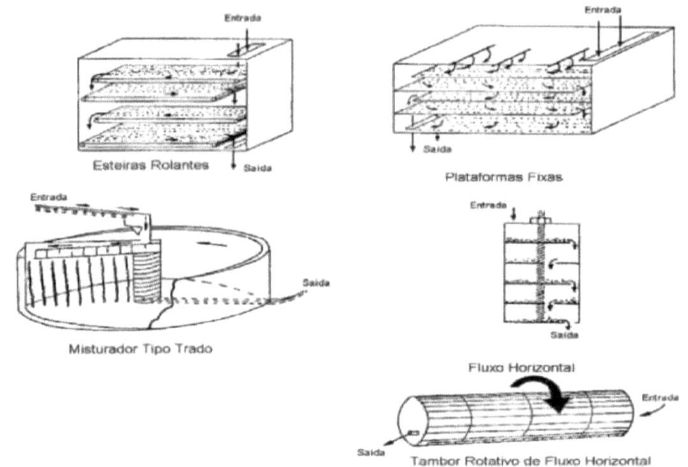

Source: Fernandes (1999)

2.5 Potential of composting in the treatment of waste

The *US EPA - United States Environmental Protection Agency* - recommends an integrated waste management system to solve the problems of solid waste generation and management at local, regional and national levels. From a holistic viewpoint, four forms of management should be considered: resource use reduction; recycling; incineration and landfill (US EPA, 1989). According to the agency, recycling should include composting. In Europe, the European Directive - EU 1999/31 - required its member countries to significantly reduce the landfilling of organic waste and composting became the main alternative to landfills, which should only receive waste (MASSUKADO, 2008).

Composting can be used in an integrated solid waste treatment , as a material recycling system or as a single system for treating organic waste. Among the advantages of this process, Russo (2003) mentions:

- The rapid microbial action on organic matter, oxidizing it and making it stable with minimal odor production;
- Hygienization of the materials being treated due to the exothermic reactions in decomposition, with good destruction of pathogenic organisms;
- The process uses very little external energy to function, compared to others, as much of the energy used comes from the metabolic process itself;
- Great flexibility in terms of scale of operation; production of organic fertilizer compounds, which do not contaminate groundwater or surface water, unlike chemical compounds. However, in order to use this compost, the appropriate legislation must be observed;
- Waste treatment system less expensive than other types, when you consider the resulting environmental gains.

Organic compost applied to soils will also offer advantages, as organic matter has a great influence on the soil's physical, chemical and biological characteristics. With regard to physical factors, organic matter reduces the apparent density, leading to a reduction in the resistance of roots to penetrate the soil, improving the structure through the formation of stable aggregates. The more agglutinated aggregates facilitate the flow of liquids and gases and the organic matter contributes to water retention. In terms of chemical properties, OM contributes to correcting the pH of acidic soils, retaining nutrients and increasing the cation exchange capacity (CEC). In terms of biological properties, organic matter provides ideal conditions for the development of micro-organisms that are beneficial to plants and inhibits those that are harmful or weeds (FERNANDES, 1999).

These advantages are realized through the proper choice of technologies, design and operation. Otherwise, poor quality compost will be formed and odors will be emitted that will negatively affect the neighborhood. The limitations of the process are that it requires more land and labor than other processes. However, the latter can be an advantage as it absorbs labor, which is almost always unskilled (RUSSO, 2003).

If the compostable material were separated at source and sent for specific treatment, the accumulation of organic matter in landfills would be avoided, making it possible to increase the useful life of the landfill or to build smaller landfills for the same period. There are also economic gains by reducing transportation costs (if composting takes place closer to the generators), final disposal and leachate treatment (MASSUKADO, 2008). The reasons for composting still being little used in municipal management programs are, for Massukado (2008, p. 44), "the difficulty of obtaining the compostable materials separated at source; insufficient maintenance of the process; prejudice against the product and little investment and adequate technology for collecting the material".

According to Web-Resol (2009), it is necessary to meet the criteria for building a composting plant:

- The existence of a consumer market for organic compost in the region;
- Efficient and regular collection service;
- Differentiated collection of household and public waste;
- Availability of sufficient area for the compost yard;
- Availability resources for initial investments, or establishment of partnerships with private groups for concession arrangements;
- Appropriate selection of technology, based on the municipality's financial availability and the amount of organic waste produced;
- Economic feasibility study, taking into account the economic and environmental benefits.

In view of the above, this work aims to analyze the potential of composting in municipal solid waste management and propose the implementation of an organic waste treatment system for the municipality of Paulista, located in the Recife Metropolitan Region, as a case study.

3. METHODOLOGY

3.1 Characterization of the study area

The municipality of Paulista is located in the state of Pernambuco (Figure 4.a), 17 km north of the capital, Recife (Figure 3. b), and is considered part of its Metropolitan Region. It is located at latitude 7°56'27" to the south and longitude 34°52'22" to the west. It has an area of 97.312 km^2, a population density of 3,087.66 inhabitants/km^2 and an estimated population of 319,769 inhabitants in 2014. It has a Municipal Human Development Index (MHDI) of 0.721 (IBGE, 2015).

The city's economy is dominated by activities linked to the service sector, commerce and industry. Tourism is also responsible for attracting businesses to the town with the establishment of hotels, restaurants, shops and marinas. It also has an industrial park that houses companies from various sectors, which boosts the region's economy. The city also has a rural area, with agro-industry as another pillar of the economy, especially the cultivation of fruit and vegetables, such as bananas, coconuts, yams, cassava, among others (PREFEITURA MUNICIPAL DE PAULISTA, 2015).

Figure 4. a) Map of the State of Pernambuco; b) Map of the location of the Municipality of Paulista - PE

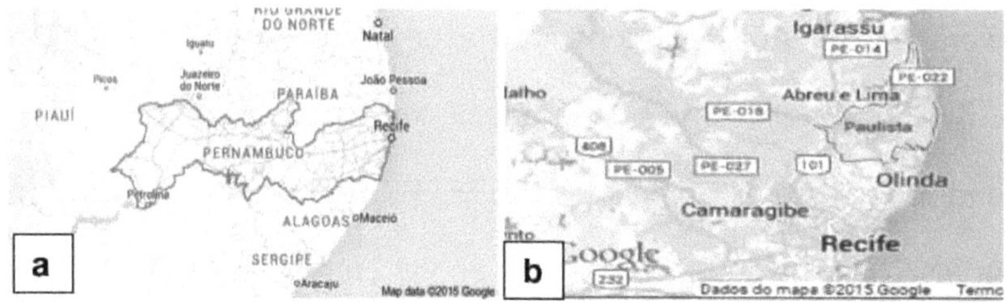

Source: Google Maps (2015)

3.2 Data collection

According to Silva and Menezes (2005), the methodology used to collect data is classified:

(i) As for its nature, it is applied, as it aims to generate knowledge for practical application;

(ii) From the point of view of the quantitative approach, this means transforming information and opinions into numbers in order to classify them;

(iii) Its objectives are descriptive, as it seeks to describe a phenomenon by collecting primary data through informal interviews and semi-structured questionnaires with the municipality's urban cleaning managers, in order to understand the current management system, as well as visits for observations and findings.

The period of visit and acquisition of primary data was from May to June 2015. Secondary data for understanding the technique and potential of composting was obtained from the literature, through books, articles and theses. The official database of the Brazilian Institute of Geography and Statistics (IBGE,

2015), the Institute of Applied Economic Research (IPEA, 2012) and the Panorama of Urban Solid Waste in Brazil of the Brazilian Association of Cleaning and Special Waste Companies (ABRELPE, 2013) were also used to support the discussion of the topic, as well as the current legislation on the subject under study, the National Solid Waste Policy.

3.3 Economic viability

After collecting the primary data, a discussion was held with the relevant literature in order to verify the feasibility of composting in the municipality's waste management, based on cost accounting, in particular, the Ministry of the Environment's Manual for Implementing Composting within Public Consortia (MMA, 2010) was followed. Once the cost data had been collected, a preliminary economic feasibility study was carried out in general, as well as an estimate of the return on investment, with a view to fulfilling the objective of evaluating the inclusion of a composting system in the municipality, taking into account local particularities.

4. PRESENTATION AND ANALYSIS OF RESULTS

4.1 MSW generation and management in the municipality of Paulista/PE

In the municipality of Paulista, collection, transportation and final disposal services are carried out through a public-private partnership. According to the data collected, the partner company that carries out these activities is obliged to comply with the municipality's Waste Management Plan and make short- and medium-term investments in measures related to achieving this Plan, in accordance with the terms of reference established for monthly payments. The cost to the municipality of this partnership is R$2,007,000.00 per month, for 25 years, according to the city's urban cleaning manager. This amount means an investment of approximately R$6.27 per inhabitant per month, which is less than the average invested in municipalities in the Northeast, which, according to the Panorama of Solid Waste (ABRELPE, 2013) is R$8.11 per inhabitant per month.

The city deactivated the dump, which was located in the Mirueira neighborhood, in 2009 and began disposing of its solid waste in a private, licensed landfill. The landfill site is in the process of being remediated due to the environmental impacts caused to the soil and groundwater, and is now a controlled landfill. Despite the restrictive legislation, 60% of Brazilian municipalities still dispose of their waste inappropriately, corresponding to around 79,000 tons of MSW per day being deposited in dumps or controlled landfills, which is not the correct destination for the waste (Figure 5) (ABRELPE, 2013). The municipality of Paulista is among the Brazilian municipalities that dispose of their waste in landfills. The cost of receiving this waste at the landfill is included in the public-private partnership.

Figure 5: Final disposal of MSW in Brazil in t/day

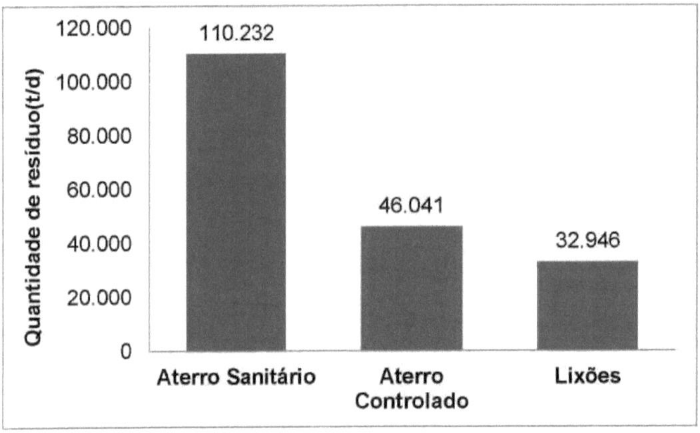

Source: Adapted from ABRELPE (2013).

According to the interviewee, the city produces 11,800 tons of solid urban waste per month. Estimating MSW generation per capita, based on IBGE's projection of the number of inhabitants for 2014, 1.23 Kg/inhab/day of MSW is generated in the municipality, which is higher than the average generation of solid urban waste in Brazil in 2013, and also in the Northeast according to a report by ABRELPE (2013) (Figure 6).

There is no selective waste collection in Paulista and organic waste is sent to the licensed landfill like any other waste. The production of organic waste is approximately 7,400 tons per month, according to data estimated by the City

Council, corresponding to 62.7% of the municipality's MSW. Despite the high percentage of organic matter, there is not, and never has been, a composting plant or other specific treatment for this waste in the municipality. However, this treatment is included in the city's Waste Management Plan and is therefore one of the obligations to be fulfilled by the private company responsible for urban cleaning.

Figure 6. Comparison of *per capita* MSW generation in Kg/inhab/day between the municipality of Paulista and the Brazilian and Northeastern averages.

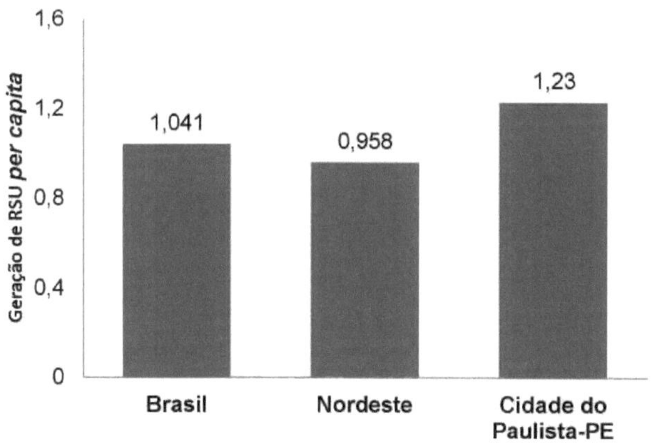

Source: Adapted from ABRELPE (2013)

In Brazil, on average, 51.4% of the MSW collected is organic waste (Figure 7), according to IBGE data (IPEA, 2012), but composting experiences in the country are still very small. Only 211 municipalities use the organic fraction of their waste for composting. Without proper collection, organic waste is sent to landfills and takes up more space, as well as forming leachate, which needs to be treated, making the operation more expensive. Of an estimated 94,309.5

t/day of organic matter collected in Brazil, only 1.6% of organic waste is sent to composting units, the rest being sent to dumps, controlled landfills and sanitary landfills, generating public health and environmental problems (IPEA, 2012).

Figure 7. Gravimetric composition of MSW collected in Brazil

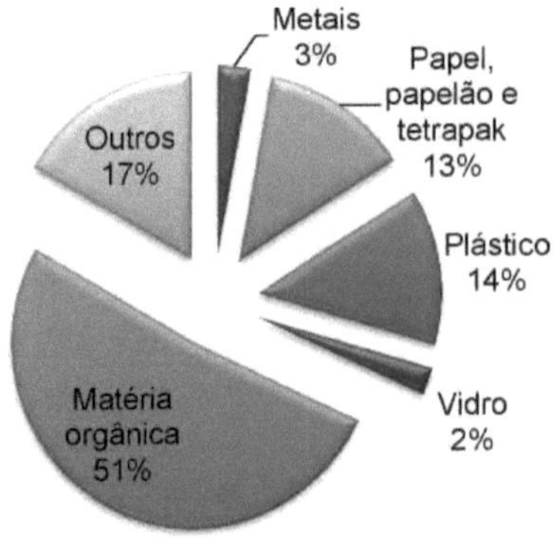

Source: IPEA (2012)

4.2 Feasibility of implementing composting in the municipality

As the city does not have selective collection of household waste, i.e. the segregation of waste into dry or recyclable material and wet material (organic waste), the feasibility of implementing composting for pruning waste, which is collected separately from the rest, was verified. According to the person in charge of urban cleaning services in the municipality, around 200 tons of pruning waste are produced every month in Paulista.

According to the Ministry of the Environment (2010), the mechanized or accelerated composting process is recommended for quantities of waste greater than 100 t/d. As Paulista's pruning waste amounts to approximately 6.6 t/d, the most suitable composting process is natural composting. Natural composting with mechanical turning of the beds using backhoes, machines that already exist in the municipality, in a covered yard to reduce the limitations of climatic variations, was analyzed.

According to the Ministry of the Environment's Manual for the Implementation of Composting (2010), the beds can be up to 2 m high. Above this height, they can compact at the bottom, making aeration difficult. They should be pyramidal or conical in shape and between 2 and 4.5 m wide. Considering a bed that is 1.2 m high and 1.2 m wide, i.e. triangular in shape, the area of the section can be calculated using Equation 1.

$As = (b \times h)/2 = (1.2 \times 1.2)/2 = 0.72\ m^2$ **Equation (1)**

Where:

As: Area of floor section

b: Width

h: Height of the bed

Assuming a waste density (d) of 550 kg/m^3 and processing of 6.6 tons of waste per day (m), it is possible to calculate the volume of the beds (V) and from the volume, the length of the beds (L).

$V = m/d = 6600kg/550\ kg/m^3 = 12.00\ m^3$ **Equation (2)**

$Length = V/As = 12.00m^{(3)}/0.72m = 16.67\ m$ **Equation (3)**

The dimensions of the compost heap are 16.67 m long, 1.2 m wide and 1.2 m high. With these values, it is possible to obtain the area of the compost yard by following equations 4, 5 and 6.

Area of the bed base (Ab):

$Ab = b \times L = 1.2 \times 16.67 = 20.004\ m^2$ **Equation (4)**

Considering the need for a clearance area (Af) for turning, equal to the area of the base, a useful area (Au) of:

Useful area (Au):

$Au = Ab + Af = 20.004\ m^2 + 20.004\ m^2 = 40,008\ m^2$ **Equation (5)**

It is necessary to take into account a further 10% of the total usable area for circulation and safety, which is 4.008 m2. Thus, the area required for each bed is 44.008 m2.

Taking into account that the average composting time is 120 days (MMA, 2010), the area of the composting yard (Ap) must be able to hold at least 120 beds at the same time, according to Equation 6:

$Ap = 44,008 * 120 = 5.281\ m^2$ **Equation (6)**

This area does not include administration or support, as well as a place to store tools and compost, and to operate a branch shredder. However, the site proposed for the composting yard in the municipality of Paulista is Mirueira, where the old dump used to be and which currently serves as the transshipment area for waste to the landfill and also the headquarters of the private company that carries out the municipality's urban cleaning services, which would be responsible for running the composting yard. Therefore, the site already has office areas, canteens and changing rooms, as well as water, sewage, electricity and telephone connections.

The area of the composting yard should be covered with metal tiles, which cost R$31.52 per m ^{2}, according to a table from the National System for Researching Construction Costs and Indices - SINAPI (2015). In addition, the unit must have a system for collecting and treating liquid effluent, which can be a septic tank with a drain or treatment ponds (MMA, 2010). In the case of Paulista, there is already a lagoon on site that can receive the leachate that may be generated in the process. Leachate occurs when there is excessive humidity in the piles of material, a fact that can happen during rainy periods or due to failures in humidity control. However, when the beds are well managed, these effluents are produced in small quantities. A concrete drainage channel installed around the yard will lead the liquids to the treatment point. According to the SINAPI table (2015), the cost of a concrete slurry drain is R$ 60.13 per meter and the composting site is about 100 m from the lagoon.

The area in which the composting yard will be located must be waterproofed with a 30 cm thick layer of compacted clay. Asphalt or concrete can also be used, with a slope of 2% in relation to the place where effluents that may be generated in the process will be collected. According to the SINAPI table (2015), the cost of mechanical compaction is R$3.05/m^3. The initial investment for building the composting yard is shown in Table 2.

Table 2. Investment for the construction of the compost yard

Description	Qty.	Unit Cost (R$)	Total Cost (R$)
Coverage	5.281 m^2	R$ 31.52/m^2	166.457,12
Compaction	1.584,30 m^3	R$ 3.05/m^3	4.832,12
Collector drain	100 m	R$ 60.13/m	6.013,00
Total			**177.302,24**

Based on the surveys carried out, the equipment and utensils needed to operate the composting technique were obtained, as described in Table 3.

market values and the Ministry of the Environment's Composting Implementation Manual (2010).

Table 3. Operational equipment required for the composting yard

Description	Qty.	Total Cost (R$)
pH meter	01	420,00
Humidity meter	01	770,00
Soil thermometer	01	218, 60
Digital Scale (up to 10 Kg)	01	89,90
20 l bucket	07	70,00
Manual sieve (8 mm mesh)	07	210,00
Pa	07	210,00
Hoe	07	140,00
Metal broom	07	105,00
50 m hose	01	120,00
Total		**2.353,50**

To process 7t/day, 7 people are needed to carry out the following activities: receiving and dispatching material, shredding branches and separating thick branches that won't be used, setting up and turning over the beds, controlling the temperature and humidity of the beds, watering the beds, cleaning the yard, sieving the compost, bagging the compost and recording the material coming in and going out (MMA, 2010).

The contractual costs involve the minimum wage plus charges, which will be considered to be approximately 60% of the wage, for a total:

Monthly contract costs= R$ (880.00 x 1.6) x 7 = R$ 9,856.00 **Equation (7)**

The cost of the raw material is considered to be zero, as the waste is used, so no variable production costs have been added. The fixed costs are the labor charges and the maintenance percentages for the building and equipment, which are 0.50% and 0.20% of the value respectively (PIRES, 2011), shown in Table 4.

Table 4. Fixed Production Costs

Description	Cost per Year
Contracts	R$ 118.272,00
Equipment maintenance	0,50%
Maintaining the structure	0,20%
Total	**R$ 128.966,60**

The result of composting is organic fertilizer and the yield from the process is around 1/3 to 1/2 of the initial volume (CERRI, 2008). Thus, in the case of the municipality of Paulista, around 200 tons per month will be processed, and considering a yield of 50%, 100 tons of organic fertilizer would be produced monthly. This can be sold by the private company responsible for the municipality's urban cleaning services and the sale price, according to market research, is between R$100 and R$150 per ton. Considering the sale price of

R$130.00 per ton of compost, this would result in a monthly income of R$13,000.00 and an annual income of R$156,000.00 from the sale of compost. There is a consumer market for the organic compost produced, as there are agricultural activities in and around the region.

The cost of treating pruning waste is currently R$127.62/t for transportation to the transshipment area, where the composting yard could be; and there is also the amount paid to the landfill company for treating this waste, which is R$60.00 per ton, according to the company's sales representative. With the implementation of the composting yard, the cost of transporting the pruning waste to the transshipment area in Mirueira will still be accounted for, but the amount paid to the licensed landfill for final disposal will be saved (Table 5).

Table 5. Cost of treating pruning waste

Description	Qty.	Cost	Total Cost (Month)
Final destination	200 t/month	R$ 60.00/t	**R$ 12.000**

Therefore, for the economic analysis, the annual revenue from the sale of organic compost would be R$ 156,000.00 and the savings from the final disposal of pruning waste would be R$ 144,000.00 (R$ 12,000 * 12) per year. Subtracting the annual fixed costs, there would be an annual cash flow of R$ 171,033.40. On the other hand, the initial investment in building the composting yard (cover, compaction and leachate collection drain) and operating equipment is R$ 179,655.74. The *payback* period will be given by dividing the initial investment by the annual cash flow (Equation 8) (Pires, 2011):

Equation (8)

Payback = Invest. Annual cash flow = 179.655,74/171. 033, 40 = ***1, 05 years***

In this way, the initial investment would be recovered in just over a year, after which there would be a profit for the Paulista municipality's public-private partnership (PPP) company. Thus, the cost charged by the PPP for the city's urban cleaning services should be reduced, representing savings for the municipal administration. In addition, there are advantages for the environment with the reduction in the volume deposited in the licensed landfill, increasing its useful life and avoiding problems with the disposal of this waste, and also the use in agriculture of an organic fertilizer to the detriment of chemical fertilizers, which are more expensive and cause environmental problems, such as the eutrophication of water bodies. Subsequently, the project could be expanded to recycle all the municipality's organic waste in composting.

5. CONCLUSION

Although the municipality does not have a composting system, like the majority of Brazilian municipalities, it should consider including this treatment, as organic waste makes up more than half of the city's MSW and could mean a reduction in costs for urban cleaning services. Treating the organic fraction would reduce the volume of waste to be transported and sent to the landfill, lowering the cost of the service and generating jobs and income, as well as contributing to the preservation and conservation of natural resources.

However, there is no selective collection, nor any differentiation between households and public areas, so the initial feasibility of composting in the municipality is with pruning waste, which is collected separately. The private company, which through a partnership with the municipality carries out urban cleaning activities and is obliged to comply with the Waste Management Plan, which includes composting, would be responsible for setting up and operating the compost bin, which would be located in Mirueira, where the company's head office is located, as well as transhipping the waste.

In order for the composting technique to become more widespread, it is also necessary to implement selective collection, making it possible to use all organic waste in this type of treatment, Equally important is the need for educational practices to make the population aware, firstly of the need to change their habits and reduce the amount of waste they produce, given the high number of MSW produced *per capita* in the municipality, and secondly of the need to adopt the practice of recycling.

Although it is not a new technique, composting has gained notoriety in recent times due to the growing concern with sustainability and because it is included in national sanitation laws and the National Solid Waste Policy, as a way of recycling organic matter. The economic, environmental and social benefits of the process contribute to the achievement of the *triple bottom line of* sustainable development, a basic premise of current public policies, with the function of the municipality offering well-being and quality of life to its citizens, who are also responsible for an efficient waste management system.

REFERENCES

AIRES, R. D.; LOPES, T. A.; BARROS, R. M.; CONEGLIAN, C. M. R.; SOBRINHO, G. D.; TONSO, S.; PELEGRINI, R. **Pyrolysis**. In: III Forum of Accounting Studies, 2003.

Only 0.17% of the solid waste collected in Recife is recycled. **G1 Pernambuco.** Published on March 23, 2012. Available at:< http://g1.globo.com/pernambuco/noticia/2012/03/apenas-017-dos-residuos-solidos- coletados-no-recife-sao-reciclados.html>. Accessed on: July 4, 2015.

BRAZILIAN ASSOCIATION OF PUBLIC CLEANING AND SPECIAL WASTE COMPANIES - ABRELPE. **Overview of solid waste in Brazil.**

2013. Available at: < http://www.abrelpe.org.br/Panorama/panorama2012.pdf>. Accessed on: May 29, 2015.

ASSOCIAÇÃO BRASILEIRA DE NORMAS TÉCNICAS - **NBR 8419**: Apresentaçâo de Projetos de Aterros Sanitârios de Residuos Sólidos Urbanos. Rio de Janeiro, ABNT, 1992.

ASSOCIAÇÃO BRASILEIRA DE NORMAS TÉCNICAS - **NBR 10004**: Residuos sólidos: classification. Rio de Janeiro, ABNT, 2004.

ASSOCIAÇAO BRASILEIRA DE NORMAS TÉCNICAS - **NBR 8849:** Apresentação de Projetos de Aterros Controlados de Residuos Sólidos Urbanos. Rio de Janeiro, ABNT, 1985.

BRAZIL. Law 12.305, which establishes the National Solid Waste Policy; amends Law 9.605, of February 12, 1998; and makes other provisions. **Official Gazette of the Union**, August 3, 2010.

CAIXETA, D.M. Generation of electricity from the incineration of urban waste: the case of Campo Grande/MS. 2005. 86 p. **Thesis** (Post-Graduation in Environmental Law and Sustainable Development) - Center for Sustainable Development, University of Brasilia, Brasilia/DF.

CAMPOS, A.L.O.; BLUNDI, C. E. Evaluation of organic matter in composting: methodology and correlations. In: Environmental management in the 21st century. Lima. **Electronic proceedings...** APIS, 1998. p 1-17.

CERRI, C.E.P. **Composting**. Sâo Paulo: Postgraduate Program in Soils and Plant Nutrition, Luiz Queiroz College of Agriculture, University of Sâo Paulo. 2008.19 p.

CORDEIRO, N.M. Composting green waste and evaluating the quality of the compost obtained: case study of algar S.A. 2010. 102 p. **Thesis** (Master's Degree in Environmental Engineering - Environmental Technologies) - Instituto Superior de Agronomia, Universidade Técnica de Lisboa, Lisbon.

COUTINHO, R. M. C.; COUTINHO, A. L. O.; CARREGARI, L. C. Incineration: a safe solution for solid waste management. In: Cleaner Production Initiatives and Challenges for a Sustainable World, 3, Sâo Paulo. **Electronic proceedings...** 2011. Available at:<http://www.advancesincleanerproduction.net/third/files/sessoes/6A/6/Coutinho_RMC%20-%20Paper%20-%206A6.pdf>. Accessed on: May 20, 2015.

FERNANDES, P. A. L. Comparative Study and Evaluation of Different Urban Solid Waste Composting Systems. 1999. 128 f. **Dissertation** (Master's Degree in Civil Engineering), Faculty of Science and Technology, Department of Civil Engineering, University of Coimbra, Coimbra.

STATE FOUNDATION FOR THE ENVIRONMENT - FEAM. **Basic guidelines for landfill operations**. Belo Horizonte, FEAM: 2006. 36 p.

GOUVEIA, J.G. Guidelines for the use of organic compost in agriculture: a proposal for municipalities with up to 100,000 inhabitants. 2012. 94 p. **Thesis** (Master's in Production Engineering) - Faculty of Engineering, Architecture and Urbanism, Methodist University of Piracicaba, Santa Bàrbara d'Oeste.

BRAZILIAN INSTITUTE OF GEOGRAPHY AND STATISTICS - IBGE. **Cities@.**

Available at:

<http://www.cidades.ibge.gov.br/painel/painel.php?lang=&codmun=261070&search=
pernambuco%7Cpaulista%7Cinfograficos:-dados-gerais-do-municipio>. Accessed on: May 12, 2015.

iNsTiTuCTo DE PEsQUISA ECoNÔMICA APLiCADA - iPEA. **Diagnosis of urban solid waste**. Brasilia, 2012.

KIEHL, E.J. **Organic Fertilizers.** Sâo Paulo: Editora Agronòmica Ceres, 1985. 492p.

MASSUKADo, L. M. Development of the composting process in a decentralized unit and proposal for free software for municipal management of solid household waste. 2008. 204 p. **Thesis** (Doctorate in Environmental Engineering Sciences) - Sâo Carlos School of Engineering, University of Sâo Paulo, Sâo Carlos.

MAZZER, C.; CAVALCANTI, O. A. Introduction to environmental waste management. **Infarma**, v. 16, n. 11/12, p. 67-77, 2004. Available at:<
http://revistas.cff.org.br/?journal=infarma&page=article&op=view&path%5B%5D=299 >. Accessed on: May 19, 2015.

MINISTRY OF THE ENVIRONMENT - MMA. **Manual for implementing composting and selective collection within public consortia.** Brasilia/DF, 2010.

"Paulista, PE". (July 15, 2015). **Google Maps. Google.** Available at:<
https://www.google.com.br/maps/@-13.4343992,-41.982566,6z>.

PEREIRA NETO, J.T.; LELIS, M.P.N. **Importance of humidity in composting: a contribution to the state of the art.** In. Brazilian Association of Sanitary and Environmental Engineering, AIDIS. Challenges for environmental sanitation in the third millennium. Rio de Janeiro: ABES, 1999. p. 1-9.

PIRES. A. B. Economic Feasibility Analysis of an Accelerated Composting System for

Municipal Solid Waste. 2011. 65 f. **Final Paper** (Degree in Environmental Engineering), Faculty of Engineering and Architecture - University of Passo Fundo, Passo Fundo.

PAULISTA CITY HALL - PMP. **Get to know the city.** Available at:< http://www.paulista.pe.gov.br/site/conheca_paulista>. Accessed on: May 12, 2015.

POLAZ, C.N.M.; TEIXEIRA, B. A. N. Sustainability indicators for municipal solid urban waste management: a study for Sao Carlos (SP). **Eng. Sanit. Ambient.**, v. 14, n. 3, p. 411-420, jul/set 2009. Available at: < http://www.scielo.br/pdf/esa/v14n3/v14n3a15.pdf>. Accessed on: May 20, 2015.

RODRIGUES, V.; CONSENZA, C. A. N.; BARROS, C. F.; KRYKHTINE, F.; FORTES, E. N. S.; Urban Solid Waste Treatment and Energy Production: Legislation Analysis for Economic Viability of Joint Solutions. **Electronic Proceedings...**XI Symposium on Excellence in Management and Technology - SEGET, Rio de Janeiro, Oct. 2014.

RUSSO, M. A. T. **Treatment of solid waste.** Coimbra: Department of Civil Engineering, Faculty of Science and Technology, University of Coimbra. 2003. 196 p.

SILVA, E. L.; MENEZES, E. M. **Metodologia da Pesquisa e Elaboração da Dissertaçâo.** 4ª ed. Florianópolis: UFSC, 2005.

National Research System for Construction Costs and Indices - SINAPI. **Inputs and Composition Report for April/2015.** Available at:< http://www.caixa.gov.br/Downloads/sinapi-a-partir-jul-2014-pe/SINAPI_Custo_ref_Composicoes_PE_052015_Desonerado.PDF>. Accessed on: July 16, 2015.

US. EPA - UNITED STATES ENVIRONMENTAL PROTECTION AGENCY, 1989. **The solid waste dilemma:** an agenda for action. U.S. Government Print Office. Washington. Available at:< nepis.epa.gov>. Accessed on: May 19, 2015.

WEB-RESOL. **Manual for Integrated Urban Solid Waste Management.** 2009. Available at:< http://www.resol.com.br/cartilha8/capitulo12b3.php>. Accessed on: May 29, 2015.

APPENDIX

APPENDIX A - Questionnaire for the manager responsible for solid urban waste in the municipality

PERSONAL DATA

1. Position

---

2. Sex: Female ( ) Male ( )

3. Technical/Academic Training

WASTE MANAGEMENT

4. Is there a Waste Management Plan for the municipality?

YES ( ) No ( )

5. Is the collection and transportation of Municipal Solid Waste (MSW) carried out by a public-private partnership? YES ( ) NO ( ) What are the terms of these partnerships with regard to the responsibilities of the private company?

---

6. What is the cost of this partnership?

---

7. Where does this waste go?

8. Do you know the amount of MSW produced in the municipality? YES ( ) NO () If yes, how much?

ORGANIC WASTE

9. Is organic waste disposed of together with other waste? YES ( ) NO ( ) If not, where is it disposed of and for what purpose?

10. Is the amount of organic waste produced in the city known?

YES ( ) NO()

If so, how much?

---

11. Is pruning and street market waste collected separately from other waste? YES ( ) NO ( )

12. Is the amount of waste produced from pruning and street markets known?

Pruning: YES ( ) NO ( ) Open Market: YES ( ) NO ( )

If so, how much?

---

13. Is the waste mentioned above disposed of together with the other waste collected?

YES ( ) NO ( ) If not, where are they destined and for what function?

14. Is there or has there ever been a composting project in the municipality? YES ( ) NO ( )

If it once existed, but doesn't currently operate, what's the problem?

---

15. Is the city's old garbage dump (Mirueira) used for another activity?

Printed by Books on Demand GmbH, Norderstedt / Germany